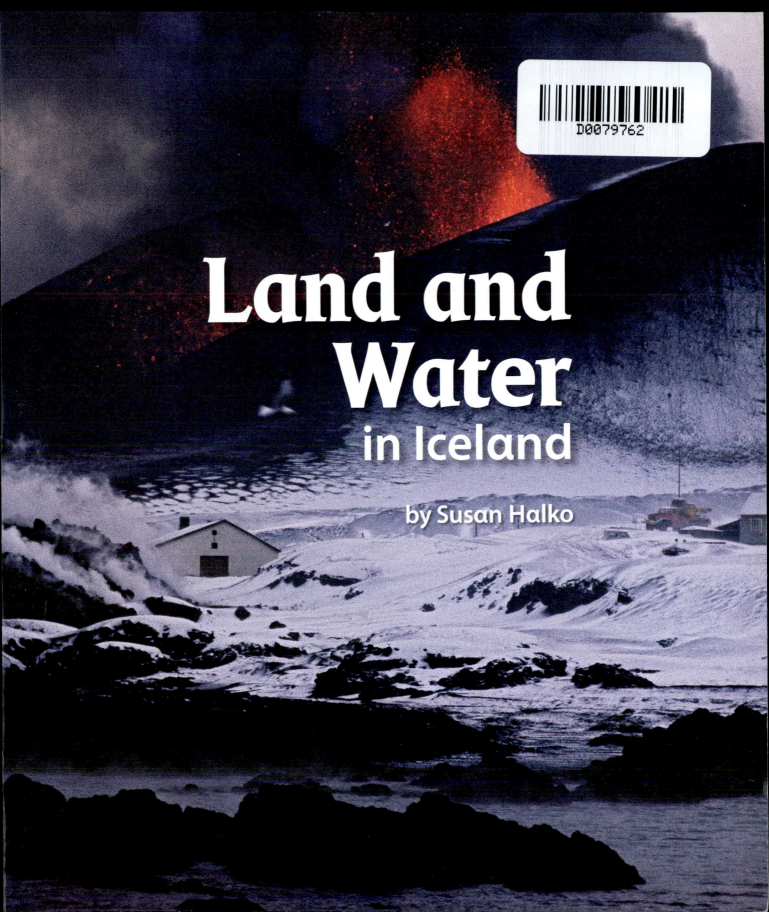

Land and Water
in Iceland

by Susan Halko

Contents

Science Vocabulary

natural resource

A **natural resource** is a part of Earth that people use.

Water is a **natural resource.** People fish in water.

ocean

The **ocean** is a large, deep body of salt water.

People eat fish that live in the **ocean** near Iceland.

mineral
A **mineral** is a nonliving material found in nature.

Basalt and other rocks are made of **minerals.**

soil
Soil is the top layer of Earth's land made of rocks, minerals, and dead plants and animals.

Farmers in Iceland grow hay in **soil.**

volcano

A **volcano** is an opening on Earth from which lava flows.

Volcanoes can cause fast changes on Earth.

earthquake

An **earthquake** is a sudden shaking of the ground caused by land moving.

An **earthquake** left this crack in the road.

weathering

Weathering is the breaking apart of rocks.

Ice cracked this rock. This is a kind of **weathering.**

erosion

Erosion is the movement of rocks or soil caused by wind, water, or ice.

Water and wind moved rocks and soil from this mountain. This is a kind of **erosion.**

Iceland

Iceland is an island. It has water all around it. Icebergs float in the water.

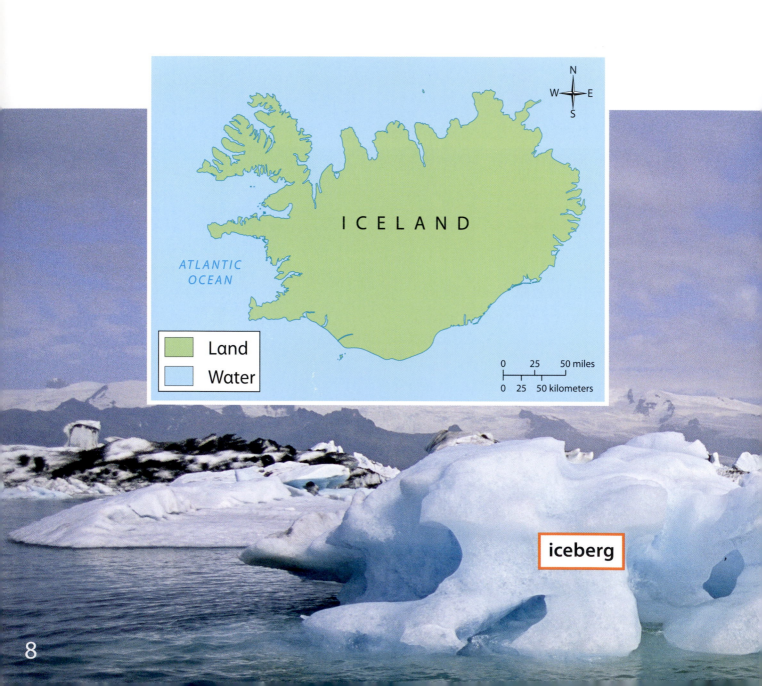

ICELAND

ATLANTIC OCEAN

N
W E
S

Land
Water

0   25   50 miles
0   25   50 kilometers

iceberg

Glaciers are huge sheets of moving ice. Iceland looks like a cold, icy place. But it has more than just ice!

glacier

Glaciers move across the land.

Iceland has lagoons. It also has lakes and rivers.

Lagoon

River

Iceland has waterfalls and grassy fields.
It has tall mountains and deep valleys.

Water falls over rocky cliffs.

People Use Water

The water around Iceland is the **ocean.**
People fish in this salty water.

ocean

The **ocean** is a large, deep body of salt water.

People in Iceland fish in fresh water, too.
Fresh water is not salty. Water is a
natural resource.

natural resource

A **natural resource** is a part of
Earth that people use.

Some water shoots up from under the ground. These are geysers.

Water boils inside the geyser.

Then it shoots into the air!

Some hot water fills pools on the land in Iceland. These are hot springs.

People cook hot dogs over a hot spring in Iceland!

People Use Land

Land is a natural resource. People in Iceland use land to raise animals.

People use wool from sheep to make clothing.

Farmers grow hay in the **soil**. They feed it to the animals.

soil

Soil is the top layer of Earth's land made of rocks, minerals, and dead plants and animals.

The Land Changes

Iceland's land changes. Some changes happen quickly. Some happen slowly.

Moving water flows over rocks. It changes the rocks over time.

A **volcano** can cause fast changes. Iceland has many volcanoes. Volcanoes can erupt on land or under the water!

A volcano erupted under the ocean. It formed this island.

volcano

A **volcano** is an opening on Earth from which lava flows.

Hot, melted rock is called lava. It shoots out of volcanoes.

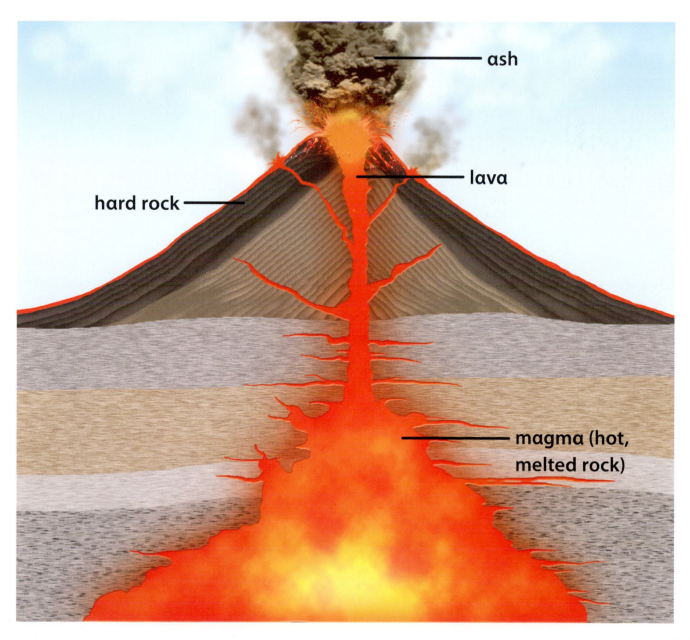

ash

lava

hard rock

magma (hot, melted rock)

Volcanoes cause changes. Some volcanoes leave behind huge holes, or craters.

This crater is filled with fresh water.

Some leave behind black rocks called basalt. A **mineral** in basalt gives these rocks their black color.

mineral

A **mineral** is a nonliving material found in nature.

An **earthquake** can cause fast changes.

An earthquake can damage roads in minutes.

earthquake

An **earthquake** is a sudden shaking of the ground caused by land moving.

An earthquake can cause slow changes, too. An earthquake split the land below this lake. Now water flows out of the lake!

Wind, water, and ice cause slow changes. Water froze in a crack in this rock. Over time, the rock split. This is a kind of **weathering**.

weathering

Weathering is the breaking apart of rocks.

Wind and water moved rock and soil from this mountain. This is a kind of **erosion**.

The land is always changing in Iceland!

erosion

Erosion is the movement of rocks and soil caused by wind, water, or ice.

Conclusion

Iceland is an island. It has water all around it. People use natural resources in Iceland. They use water, rocks, and soil.

The land changes. Some changes happen quickly. Some changes happen slowly.

Think About the Big Ideas

1. How do people in Iceland use land?
2. How do people in Iceland use water?
3. How does the land in Iceland change?

Share and Compare

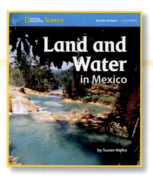

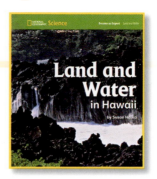

Turn and Talk

Compare ways people use fresh water and salt water. How are they the same? How are they different?

Read

Read your favorite page to a classmate.

Write

Write about your favorite photo in this book. Then share your writing with a classmate.

Draw

Draw three ways you use natural resources. Show your drawing to a classmate.

Meet Katey Walter Anthony

Katey Walter Anthony is a scientist. She learns by observing. Katey observes land and water in the Arctic. The Arctic is a cold place. It has ice and snow.

The ice is melting. The land is warming, too. A gas goes into the air when this happens. Katey wants to learn how to use this gas.

Index

Acknowledgments
Grateful acknowledgment is given to the authors, artists, photographers, museums, publishers, and agents for permission to reprint copyrighted material. Every effort has been made to secure the appropriate permission. If any omissions have been made or if corrections are required, please contact the Publisher.

Photographic Credits
Cover (bg) Jonas Bendiksen/National Geographic Image Collection; Cvr Flap (t), 5 (b), 17 Frantisek Staud/Alamy Images; Cvr Flap (c), 4 (b), 12 James A. Sugar/National Geographic Image Collection; Cvr Flap (b), 18-19, 28 Photopat iceland/Alamy Images; Title (bg) Emory Kristof/National Geographic Image Collection; 2-3 Sisse Brimberg/National Geographic Image Collection; 4 (t), 13 Johann Helgason/Alamy Images; 5 (t), 23 Vigfus Birgisson/Nordic Photos/Getty Images; 6 (t), 20 Photo Archive Submitter/National Geographic Image Collection; 6 (b), 24 Arctic-Images/Corbis; 7 (t), 26 Fritz Polking/Peter Arnold, Inc.; 7 (b), 27 Arco Images GmbH/Alamy Images; 8-9 (bg) Sisse Brimberg/National Geographic Image Collection; 9 (inset) Lauri Dammert/Pictorium/Alamy Images; 10 (t) June Morrissey/Alamy Images, (b) David Newham/Alamy Images; 11 Sisse Brimberg/National Geographic Image Collection; 14 (l) blickwinkel/Lohmann/Alamy Images, (r) Emory Kristof/National Geographic Image Collection; 15 Jonas Bendiksen/National Geographic Image Collection; 16 James A. Sugar/National Geographic Image Collection; 22 Jorge Bai/Alamy Images; 25 Haraldur Stefansson/Alamy Images; 30, 31 Institute of Northern Engineering, University of Alaska, Fairbanks; Inside Back Cover (bg) Emory Kristof/National Geographic Image Collection.

Illustrator Credits
8 Mapping Specialists; 21 Paul Mirocha

Neither the Publisher nor the authors shall be liable for any damage that may be caused or sustained or result from conducting any of the activities in this publication without specifically following instructions, undertaking the activities without proper supervision, or failing to comply with the cautions contained herein.

Program Authors
Kathy Cabe Trundle, Ph.D., Associate Professor of Early Childhood Science Education, The Ohio State University, Columbus, Ohio; Randy Bell, Ph.D., Associate Professor of Science Education, University of Virginia, Charlottesville, Virginia; Malcolm B. Butler, Ph.D., Associate Professor of Science Education, University of South Florida, St. Petersburg, Florida; Nell K. Duke, Ed.D., Co-Director of the Literacy Achievement Research Center and Professor of Teacher Education and Educational Psychology, Michigan State University, East Lansing, Michigan; Judith Sweeney Lederman, Ph.D., Director of Teacher Education and Associate Professor of Science Education, Department of Mathematics and Science Education, Illinois Institute of Technology, Chicago, Illinois; David W. Moore, Ph.D., Professor of Education, College of Teacher Education and Leadership, Arizona State University, Tempe, Arizona

The National Geographic Society
John M. Fahey, Jr., President & Chief Executive Officer
Gilbert M. Grosvenor, Chairman of the Board

National Geographic School Publishing
Hampton-Brown
www.NGSP.com

Printed in the USA.
RR Donnelley, Menasha, WI

ISBN: 978-0-7362-5519-6

16 17 18 19 20 21 22

10 9 8 7 6 5 4 3